BOO

19 projects you can do at home

Jo Scheer

Illustrations by Andrew Schaper

Photographs by Richard G. B. Hanson II, Jo Scheer and Paul Avis

First Edition
08 07 06 05 04 5 4 3 2 1

Every effort has been made to ensure all the information in this book is
accurate; however, due to differing conditions, tools, and individual skills,
the publisher cannot be responsible for the outcome of your project.

The publisher and author encourage you to take all safety precautions that
would be usual when working with shop tools and materials. They bear no
responsibility for any injuries, losses, and other damages that may result
from the use of the information in this book.

Published by
Gibbs Smith, Publisher
P.O. Box 667
Layton, Utah 84041

Orders: 1.800.748.5439
www.gibbs-smith.com

Designed by Kurt Wahlner
Produced by Steven Rachwal Design
Printed and bound in Hong Kong

Library of Congress Control Number: 2004108297
ISBN 1-58685-220-5

ACKNOWLEDGEMENTS

*I would like to thank the staff of the Tropical Agriculture Research
Station in Mayaguez, Puerto Rico, for their professional assistance in
procuring and identifying the bamboo at the station. The huge selection
and variety of bamboo available to me greatly extended my creative
horizons and enhanced my personal bamboo appreciation and experience.*

Dedicated to Laura, Sam, & Sophie, who have tolerated my bamboo predilection.

CONTENTS

INTRODUCTION

Bamboo grows on you. It will grab your consciousness and there will be no turning back. Once awareness has been ignited, bamboo will become part of your life. For those lucky enough to have bamboo a part of their natural world, it is a no-brainer. Bamboo will serve them in many ways—in play, in work and in life.

Bamboo is user-friendly, low-tech and easy to work with. Artisans and rural folk in the East have known this for millennia. My experience with bamboo spans twenty years, and bamboo projects have brought me immense satisfaction. With basic tools, a natural material and a synergy of hand and mind, all things are possible—some of which are detailed in the pages that follow. Hopefully, this will be only the beginning, with the techniques and ideas serving as platforms for yet more.

Bamboo, as a raw material or as a prop for life, fits into a much broader philosophy. The incorporation of bamboo into your life fits a lifestyle that reflects an awareness of the natural world, our place within it and a consequent behavior that exemplifies the relationship. We become aware of the necessity to live within the natural constraints of nature—to live in harmony, to live sustainably, and to be reverent.

In the East, bamboo has been revered for centuries for its beauty, utility, resilience and simplicity. Bamboo is all these things, and more. It is a gift to man. Work with it for fun, make something that gives pleasure and you will realize the gift. Its employment for the most pedestrian of uses elevates the use to something more. A fence? Commonplace. A bamboo fence? Art. Utility, though an apt application of bamboo, is not the real achievement. It is how the bamboo does its job. Bamboo makes the ordinary extraordinary. A bamboo anything is special. Perhaps we recognize the personal human input, to fashion this thing from the raw material. We see how the bamboo yielded to this design, and we see how it is so appropriate and efficient.

GETTING STARTED

Bamboo can be worked into anything—from towering tree houses to toothpicks. It is not much fun to make toothpicks, and a tree house, though a very worthy endeavor, is a large project. Fortunately, most projects with bamboo are of a scale that they can be completed in one day or less. Inspiration can come at any time, although collecting tools and materials may require a more extended time period. Once set up, however, a project can be worked through the day and finished in time for reflection of the day well spent.

First, you will need bamboo. There are a number of retail outlets around the country that are more than willing to sell you bamboo of all shapes and sizes. A quick and easy resource is the American Bamboo Society Web page, *www. abs.org*. They have a list of retailers, growers and importers, as well as a slew of information. The bimonthly magazine lists the membership every year, and you can find members that live in your vicinity. Communication with these often knowledgeable people may elicit a source for your bamboo project. They often know of local bamboo stands or know of someone else that does. Another source is nurseries or even botanical gardens. And, upon acquiring permission to cull some bamboo, the actual harvesting must be done in a clean and efficient manner. The bamboo will benefit from removal of dead bamboo culms, which are often the best, as they are cured and have low moisture content. The stand must be left neat and tidy, or you may not have permission to return. The American Bamboo Society is another place to find bamboo aficionados.

Secondly, you will need tools. If you already have a history of woodworking, you will have acquired an inventory of tools easily adaptable to working with bamboo. Some additional tools may be needed and are best to have before you start.

TOOLS

For smooth, fast, precise and loud bamboo shaping, you can't beat a good power tool. The alternative is muscle power, which has its place and is definitely quieter. Hand tools will be best for quick and specific shaping.

Belt Sander: I like the belt sander. It smoothes rough edges quickly. It can be used to fine tune an angle cut or make an otherwise difficult shape. I use cylindrical belt sanders to smooth out holes or adjust them to the eccentricity of a particular piece of bamboo to be inserted.

Brass or Galvanized Wire, 18 to 20 gauge: To prevent splits from continuing through a node, I wrap wire around the culm in two to three loops. Depending on the piece and the use, I use either brass or galvanized wire, 18 to 20 gauge.

Cordless Drill: The cordless drill, with a full complement of hole saw sizes, is something I use often.

Power Compound Mitre Saw: I am enamored with the power compound mitre saw and its 80-tooth, 12" blade that easily cuts through bamboo. I like the precise angles possible and the tight fits that can be achieved. With mitred bamboo, an otherwise difficult joint can be done with ease.

Router: I love my router, and use it to round over a bamboo edge.

Saw: I always carry a foldable pruning saw. The precision-ground blade is very effective at cutting bamboo and is easily worked into tight spaces. Smaller tooth size helps but is not essential.

Spoke Shaver: Although it's heavy, I like my spoke shaver. The tool is used to shave off bark from trees and is very effective at splitting large culms and working your way through several nodes. Pound it through and twist—it works great. Alternatively, I have used a machete, but I prefer the spoke shaver.

Swiss Army Knife: First and foremost, and without a thought to promoting an otherwise proprietary product, I absolutely need my Swiss Army knife. As there are several, let me list the essential blades and attachments. Besides the large and small blades, which are very good at doing the intricate bamboo splits, my favorite blade is the saw. The arrangement of serrations makes the saw very effective in cutting precise incisions in bamboo. It is also effective at slight edge alterations—from holes to mitres. The saw can also be used to auger a hole up to about $1/2$" in diameter, if initiated by another tool, the leather punch, also used as an auger.

Vise Grips: An essential tool for grasping the wire ends and twisting to achieve a tightly wound bind. It can lock on the wire with super-human grip, allowing a twist to tighten the loops.

Wire Cutters: A wire cutter is used to cut off the excess wire that you usually use to prevent splits from traveling too far.

Glue: A good wood glue, preferably one that dries clear, works best.

BAMBOO ANATOMY TERMS

Branch Scar: The point on the node where a branch would emerge. Often, lower nodes do not have branches, only the unemerged branch scars. These are alternately arranged on proximal nodes (one side, then the other).

Culm: A long tube with regularly spaced membranes, or node septums, designed to prevent collapse of the tube under undue lateral stress.

Internode: The smooth tube between the nodes. The lower internodes are thicker and shorter, gradually progressing to longer internodes with thinner walls as you go up the culm.

Lignin: An interconnecting fiber that is slowly deposited in the wood of bamboo, making it stronger in tensile strength. It is this deposition that makes culms older than about four to five years a better choice for harvesting.

Node: A visible seam circumscribing the culm, delineating a solid membrane that traverses the internal void. This is the growth point. The straight grain of bamboo gets confused here, and it is less likely to split.

Septum: The brittle and easily removed membrane across the interior of the culm at the node.

Skin: The outside surface of the bamboo—hard and thin.

Wall: The full depth of the culm "wood," varying in thickness according to species and culm section. The wall thickness is a variable dependent on species and height of the culm internode. Lower wall thickness is greater, getting thinner the higher up the culm you go. For the purposes of this book, thick-walled bamboo averages $5/16$" to $3/4$", while thin-walled bamboo is $3/16$" to $1/4$".

TIPS FOR
WORKING WITH BAMBOO

What makes bamboo so great to work with? First, it has a natural finish. The <u>culm</u>, with its hardened, silicon-rich skin, needs only to be cleaned and polished. The patina of the surface varies with species, but most are very pleasing to the touch and to the eye. Bamboo is round, generally hollow—a tube with periodic nodes that are solid. As a lightweight structural member, it has no equal.

Splitting: Bamboo grain lends itself to perfect, precise, straight splits. This can be used for many applications, the highest form of which is exemplified by the Japanese tea whisk. Bamboo is flexible: it has a high tensile strength as well as compressive strength. It can be distorted to a great degree, yet maintain the ability to return to its original shape. Split bamboo allows even more distortion and is the basis for many other applications, perhaps the most famous being the bamboo fishing pole. Whatever your particular project, recognition of bamboo split character and management is essential.

1. The node is a natural barrier to the progression of a split. If a short split is desired, the node is always the terminus. As insurance, I generally wrap the bamboo, either at the node or between the node and the splits, with 18- to 20-gauge brass or galvanized wire. Two to three loops usually allow an easy cinching. The wires are twisted together, making sure there is only one overlap of the loops. Using a Vise Grips, the wire can be grasped, pulled and twisted to tighten. The excess wire is cut off with a wire cutter, and the end is pushed neatly against the bamboo culm. The wire can fail if twisted too much or if not twisted evenly. I have used twine, cord, string and stretchable electrical tape. The choice depends on function and aesthetics. Your call.

2. When a design calls for a split through a node, a little extra effort in splaying the split will generally work. However, due to a confusion of grain within the node, the split may not emerge in perfect alignment with the original split. If the splits are few, perfection is not an issue. If splits are many, adjacent splits can merge, with resulting frustration. Careful node splits with a knife can reduce this possibility, but generally many splits through a node are not advised.

3. When using whole culms of any length longer than a couple of internodes, the tendency to split needs to be reckoned with. When I make floor lamps, the top and bottom of the whole culm is bound in wire, preventing the progression of the splits at the base and the shade. If a particular application allows, a pre-split whole culm will prevent the creation of any additional splits that are due to rapid changes in temperature and humidity. A design that incorporates splits is an effective strategy against them.

4. Splits should always be made by splitting the culm, or culm section, in half. This technique can be done progressively up to sixty-four splits around the culm. I have done 128 with larger-diameter and straight-grain bamboos *(B. tulda)*. If splits are made off center of half, they tend to progress towards the thinner side, which accentuates the imperfection. Therefore, care in making the first incision with a knife into the end grain cannot be emphasized enough when making many splits, as in a lamp shade. However, if a split does not work out, the thinner one can be removed, with a yank, and will not be missed by even the most scrutinizing observer. It will be your secret.

5. Splits should be made such that the direction is always through the center of the culm, the shortest distance across the outer wall. I have done off-center splits, taking a section off the culm. Again, the split tends toward reducing the smaller split-off section. It is manageable, but an issue.

Safety: I would be remiss if I failed to address a character of bamboo that may raise its ugly head most unexpectedly. As bamboo splits very cleanly, it also splits very sharply. The sharp edges can cut the skin, and care must be taken to avoid a run for bandages. I have run a blade at a right angle along the edge of concern. The blade will remove a sliver of edge and thus round it off.

Joinery: The tube shape of bamboo restricts the variety of joinery techniques. Much research has been devoted to joinery with emphasis on strength, as the joint is generally the weakest link in a bamboo structure. However, good design can alleviate much of the strength requirement and allow simpler joinery techniques.

1. A fundamental architectural rule of structural integrity is triangulation. A triangle prevents distortion and is a good technique to prevent stress at joinery. With triangulation, bamboo can be mitred. If cuts are made precisely on the same culm, with the same orientation, the mitre approaches perfection in surface irregularities—both legs of the mitre have the same imperfections and are matched. With precise cuts, the end grain of the bamboo can be matched and glued. The difficulty of setting up a jig to press the joint together while the glue sets has led me to use electrical tape. The tape stretches, exerts substantial pressure on the joint, and allows it to set.

2. The mortise and tenon is probably the most used joinery technique. It is simple but requires the diameter of the tenon bamboo to be less than the interior diameter of the mortise bamboo. With a hole saw and some whittling, the joint can be done quickly. A dowel drilled through the side of the bamboo joint to the other side secures it. The joint is not rigid. Subsequent rigidity must again be accomplished by triangulation. Triangulation can be done with wire, under tension, in some circumstances.

3. Another effective joint is interwoven bamboo splits, oriented end to end. This is used in a variety of applications, many as split bamboo lamp shades. The bamboo ends are split into equal numbers, either 8, 16, 32, or 64, and interlaced.

4. Smaller-diameter bamboo pieces can be wedged into the cavity of a larger bamboo. Again, the smaller bamboo is split, and a bamboo ring is pushed into the end to splay out the splits. These splits will compress into the larger bamboo. With glue, it can be made permanent. This is one technique for inserting a lamp socket holder into a split bamboo shade.

5. A general rule, with exceptions, is the need to keep all joinery 1:1. A joint can join two pieces, coming from different directions, but it is very difficult to make a decent joint for three bamboo pieces. The third should be designed to join at some other point, as a mortise and tenon.

Attachments:
Because of its tendency to split, bamboo should not be nailed unless pre-drilled with a similar-diameter drill bit. Screws and dowels require pre-drilling as well. It helps to have a full set of drill bits that accurately match the various sizes of hardware.

1. Whole culms can collapse when a screw or nail compresses it. Being careful not to compress a screw or nail is one preventive technique. If a hollow bamboo dowel is fitted into the larger bamboo from the opposite side, a much tighter joint can be accomplished, as the dowel compresses against the outer bamboo wall.

2. Butt joints of bamboo to wood require an internal wood or bamboo dowel that fits the internal diameter of the bamboo. This is cut short enough such that it can be attached to the wood surface with a long screw. The bamboo is then secured to the dowel. Very tight joints can be accomplished if the dowel screw is from the backside of the wood surface. Attach the bamboo to the dowel, then tighten down on the dowel.

Bamboo Holes: The tendency of bamboo to split is both an exploitable design advantage and an unavoidable construction disadvantage. Splits, when desired, are easily achieved with a technique respective of bamboo morphology. A design that exploits bamboo splitting is fundamental to good bamboo design. But there are times when a split is not desired. Most bamboos will split when a nail is driven through the skin. The nail acts as a wedge to pull the straight grain of bamboo apart. Consequently, drilling is required.

1. Normal drill bits can work fine, assuming they are sharp. However, a slow and gradual technique is desirable to prevent a sudden penetration of the bamboo wall, which is not very thick and easy to go through. I have used Forstner drill bits, and they work great. They cut the outer edge of the hole first, making a clean cut.

2. A hole saw works in a very similar manner, but some precautions are necessary. The central guide hole must be made slowly and carefully. This prevents breaking through the wall suddenly and subsequently slamming the hole saw teeth on the bamboo skin. Once the wall is penetrated with the guide bit, the hole saw is gently started on the bamboo. The hole is gradually completed, and once again care must be taken to prevent a sudden breakthrough. Once through, the bamboo hole must be removed from the hole saw bit. It may be necessary to remove the central guide bit once the hole saw has established a groove. This prevents the guide bit from penetrating the opposite sidewall of the bamboo. However, if the hole is intended to go completely through the bamboo, the guide hole can be used to start the hole from the opposite side. A hole pushed through from the inside out will splinter and is to be avoided.

3. Most holes are made at a right angle and are easily made. However, some designs may require a 45-degree angle. A normal bit can be started at a right angle, then rocked back to a 45-degree angle to complete the hole. The hole saw is done in a similar manner.

FUN & FUNCTIONAL

CANDLE LAMP
EASY

Perfect for an outdoor party, or as a dining room centerpiece,

this candleholder is easy to make. Summer flowers add scented delights!

Materials

- *Thin-walled bamboo 3"-5" in diameter*
- *18- to 20-gauge Brass wire*
- *Glass candleholder*
- *Candle*
- *Flowers*

Tools

- *Knife*
- *Pencil*

Step One: Cut a piece of bamboo at a 90-degree angle just below a node (the section where two pieces of bamboo are joined), and again 10" above the node (fig. 1). This is the top of your bamboo.

Step Two: Draw a ring around the culm 3" above the node (fig. 1).

Step Three: Make an incision into the top end of the bamboo. The knife will initiate a split. Split the bamboo until it reaches the drawn circle. Make a second split on the opposite side of the stalk (fig. 3).

Step Four: Repeat splitting into $1/4$" sections, $1/8$" sections, $1/16$" sections, and so on until the width of the bamboo between the splits is about $1/4$" wide—like slicing a pizza (fig. 2).

Step Five: Pull out every other section of split bamboo at the pencil mark (fig. 1).

Step Six: Fit a glass candleholder within your bamboo, and place a candle in its base (fig. 2).

Step Seven (optional): For an extra touch, arrange flowers within the bamboo slots.

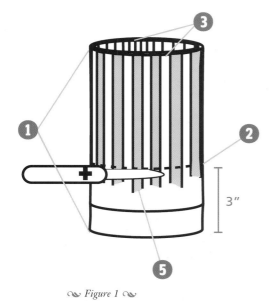

∾ *Figure 1* ∾

3"

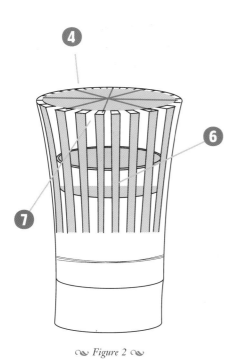

∾ *Figure 2* ∾

PICTURE FRAME
EASY

Whether framing a work of art or a favorite photo, bamboo's rugged texture and neutral colors will accent most pictures.

Materials

- ❧ *2 bamboo culms 10" long*
- ❧ *2 bamboo culms 8" long*
- ❧ *6 feet of twine*
- ❧ *4 bamboo branch segments 2" long*

Tools

- ❧ *Saw*
- ❧ *Drill and Drill Bit Set*
- ❧ *Hole Saw Kit*

Step One: Select your materials and cut the bamboo to the specified dimensions for an 8" x 10" picture frame.

Step Two: With the hole saw, cut the holes by going through the backside of the bamboo with the guide bit to receive the horizontal bamboo pieces as in figure 1.

Step Three: With a drill, make a hole complementary to the twine diameter at the inside corner of the joints (fig. 2).

Step Four: Cut four, 2" sections of small-diameter branch bamboo (fig. 4).

Step Five: Assemble the frame. Thread the twine as shown in figure 3, and continue through the outside hole. Tie each emerging twine as tight as possible to the center of one bamboo twig.

Step Six: To adjust and tighten, rotate the bamboo twig segments by twisting the twine (fig. 4).

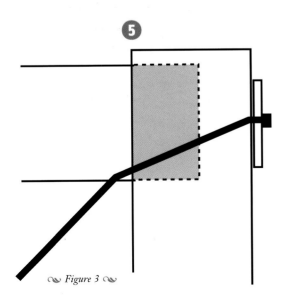

❧ Figure 1 ❧

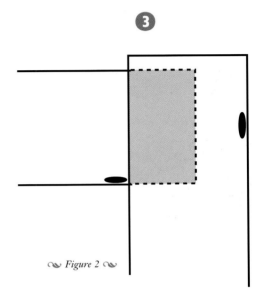

❧ Figure 2 ❧

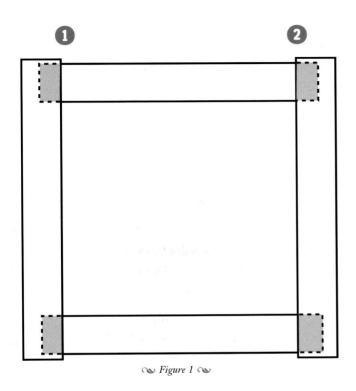

❧ Figure 3 ❧

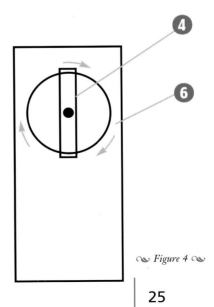

❧ Figure 4 ❧

TREASURE BOX
MEDIUM

It is not actually a box, in that it isn't square, but it will hold all your treasures and make a perfect place for jewelry.

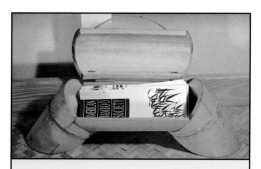

Materials

- ❧ *12"-long, large-diameter bamboo with three internodes*
- ❧ *3" bendable piece of a bamboo branch*
- ❧ *Electrician tape*
- ❧ *Small Hinge*

Tools

- ❧ *Power Mitre Box*
- ❧ *Wood Glue*
- ❧ *Hacksaw*
- ❧ *Knife*
- ❧ *Screwdriver*
- ❧ *Drill and Drill Bit Set*

Step One: Using a power mitre box, cut both sides of your 12" bamboo with equal 45-degree-angle cuts that do not traverse the inner node (fig. 1). These cut pieces are your "cut-offs."

Step Two: Rotate cut-offs 180 degrees, and align them to the central internode cut that the cut-off came from (fig. 2). If the alignment does not pass your inspection, you can again cut a 45-degree angle from the cut off, though exactly opposite of the first cut (fig. 3). This orientation is assured to be very close to the central piece in surface line-up.

Step Three: Settled with your attachment strategy, cut the cut-off at 1" to 2" beyond the 45-degree cut, with a 90-degree cut. Assemble the cut-offs and the central internode. If you are very lucky, the box will sit flat on the 90-degree cut-offs and the mitre will line up satisfactorily. You can then glue the pieces, binding the mitres with electricians tape and making sure it still sits flat while the glue sets (fig. 4). If you are not so lucky, glue the cut-offs symmetrically and bind with electricians tape. When the glue has dried, set the box on a surface and mark where and how you need to adjust the cut so it rests perfectly. This is a progressive technique requiring a fine tuning of each surface to achieve stability. The cuts can be made with the mitre saw if done carefully.

Step Four: Cut the top door of your box (see photo) carefully with the hacksaw. Read step five, then mark your side cuts parallel to each other. Also, slant the blade toward the center slightly when making the cut. This precaution is insurance that the door will not bind when opening and closing. Mark the ends of these side cuts such that the door envisioned is large enough for access (fig. 5).

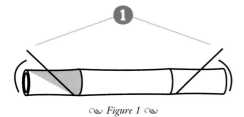

Figure 1

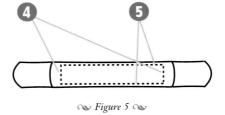

Figure 5

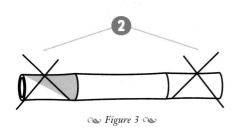

Figure 2

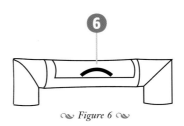

Figure 6

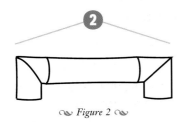

Figure 3

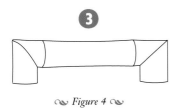

Figure 4

Step Five: Once the sides are cut, work a knife blade into the bamboo along the grain and adjacent to the termination of the side cut. Gradually work the blade in, and twist to get the bamboo to split. It should end up very close to the termination of the opposite side cut (fig. 5). If the opposite cut is a little short, you can easily extend it. If too far, you will have to live with it. If you are really good, you will cut one side purposely shorter, with the intention of finishing the cut when you know where the split will end up. Finally, remove your door and test it for easy opening and closing.

Step Six: Screw your hinge in place.

Step Seven: To make a handle for the door, drill holes to accept the ends of the branch at the anticipated angle and at the minimal distance apart that the bamboo branch will bend. Insert, bend, then insert the other end (fig. 6).

Note: The picture shows the legs at an angle. This is an optional way of constructing the box.

WIND CATCHER

The cupped shape of half-sections of bamboo is a natural for catching the wind, and it is mesmerizing to watch as it spins.

Materials

- ᗊ *12" of the largest-diameter bamboo you can find with internodes at both ends*
- ᗊ *18" of bamboo about 20 percent smaller in diameter than your 12" section and with node at one end*
- ᗊ *6' of galvanized or brass wire*
- ᗊ *6' of fishing line*
- ᗊ *Fishing swivel*

Tools

- ᗊ *Saw*
- ᗊ *Wire Cutter*
- ᗊ *Drill and Drill Bit Set*
- ᗊ *Scissors*

The physics of converting the movement of air to the spinning of an object is important. Some principles should be kept in mind:

- The farther out from the center of rotation the wind-catching cup is placed, the more torque is applied and spin will be gained with less wind.
- The bigger the wind-catching surface, the more force is gained by air movement.
- Increasing the number of wind-catching blades increases how much wind it will catch.

Step One: With galvanized or brass wire, make two to three loops around the thin bamboo just below the upper node (fig. 3). Cut excess wire.

Step Two: Split and remove two sections of the thin bamboo, on opposite sides of the culm, of a width large enough to accommodate the culm thickness of the larger bamboo. The sections can be scored with a saw blade just below the wire loops and then broken off by bending outward (fig. 1). (See page 16 for splitting techniques.)

Step Three: Cut the large bamboo in half lengthwise. Insert the large bamboo halves into the slots made in the thin bamboo with the cups of the bamboo sections facing in opposite directions. Ideally, the inserted bamboo halves will overlap inside the thinner bamboo (fig. 2).

Step Four: With galvanized or brass wire, make two to three loops around the thin bamboo just below the protruding culm halves (fig. 3). This will secure the integrity of the piece. Cut excess wire.

Step Five: Drill through the thinner bamboo at the upper node and string with fishing line. Then tie the line to a fishing swivel. The whole piece may then be suspended by another string from the swivel to your desired location (fig. 3).

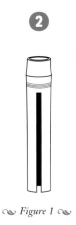

Figure 1

Figure 2

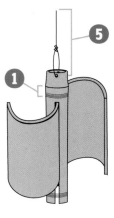

Figure 3

RAIN STICK
EASY

When tipped slowly end over end, the rain stick mimics the gentle sound of falling raindrops. Some even say that it helps bring showers during drought. The secret of its inner workings has remained a mystery to all but a select few. The mystery is part of its magic, though, so remain silent when mystified listeners ask, "But how?"

Materials

- *36"-long, medium-diameter bamboo culm with a node on both ends*
- *1 separate piece of bamboo cut just longer than the diameter of your 36" bamboo*
- *1 bandanna*
- *18- to 20-gauge wire (36" to 48")*
- *1 cup of ⅛"-diameter pebbles or dried peas*

Tools

- *Saw*
- *Drill and Drill Bit Set*
- *Hammer*
- *Knife*
- *Wood Glue*
- *Wire Cutter*

Step One: Knock out the internode septums of the 36" culm. You do not need to knock out the complete septum, but enough to allow pebbles to gently pass through; the more obstacles during the travel of the pebbles down the length of the stick, the longer the sounds of rain lasts.

Step Two: Pull off thin splits from your internode section. (See splitting technique on page 16.) Cut the splits at a width that is half the wall's thickness and at a length that is the diameter of the bamboo plus ½". If you then split the width dimension, you will have two square splits (fig. 1). These will be your pegs.

Step Three: The idea here is to fit these square pegs into a round hole, which will provide obstacles for your pebbles. Drill holes of the proper diameter into the rain stick. Proceed in a spiral fashion, with a rotation of about 1" between holes on the circumference and about 1" farther down the stick (fig. 2). This is subjective, obviously. Drill at an angle perpendicular to the bamboo surface.

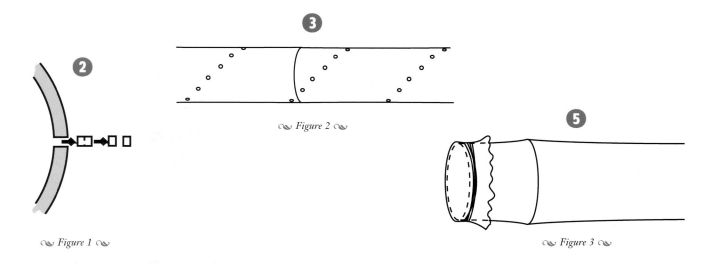

②

❧ Figure 1 ❧

③

❧ Figure 2 ❧

⑤

❧ Figure 3 ❧

Step Four: Once all holes are drilled through the complete length of the bamboo, pound a peg through each hole until it butts against the opposite side. Cut off excess peg length with a sharp knife.

Step Five: The ends of the stick need to be sealed tight. Do one end first before the rain stick is filled with pebbles or peas. Place a piece of cardboard over the end, and then trace the outline of the bamboo onto it. Cut out the cardboard along the mark, place it over the bamboo end and glue it in place. Place the bandanna over the cardboard. Pull it tight while wrapping two or three loops of wire around the material and the bamboo. Twist the wire tight, tucking the ends under to make them secure. Cut off excess wire and material (fig. 3).

Step Six: Place 1 to 1¹/₂ cups of peas or pebbles into the rain stick. Seal the remaining end the same way as you did the first end. Gently tip, twist or shake the rain stick to coax out the soothing, relaxing and stress-relieving sounds of rain.

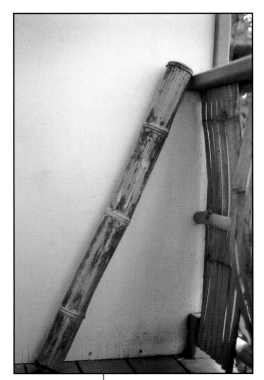

EARRINGS
EASY

Materials

- ∽ 1"-long, ¼"-diameter bamboo twigs
- ∽ 20-gauge jewelry wire, silver or brass (5')
- ∽ 2 head pins (2") and ear wires (found at jewelry-making and bead stores)
- ∽ 8 beads (found at jewelry-making and bead stores)

Tools

- ∽ Knife
- ∽ Cutting Board
- ∽ Candle or Butane Lighter
- ∽ Flat-Nosed Jewelry Pliers
- ∽ Round-Nosed Jewelry Pliers
- ∽ Wire Cutter

Venezuelan-style earrings and a matching necklace make charming gifts for any occasion.

Step One: To make the four bamboo, oblong beads, cut the bamboo twigs with a sharp knife on a cutting board. Press the knife into the bamboo with a rolling motion. Continue this until the end is cleanly cut through.

Step Two: Burn all the bamboo ends by passing them over a candle while holding the bamboo piece with flat-nosed pliers.

Step Three: Create the base for the design to hang from (see diagram). With round-nosed pliers, grasp the end of the wire. Wrap the wire around one nose of the pliers, then the other, in a succession of double-stranded circles. Keep the loops similar in size and wrap tightly. Using flat-nosed pliers, flatten each loop piece individually. Match pairs for earrings.

Step Four: Thread your bead/bamboo combination onto a 2"-head pin.

Step Five: Cut off excess wire, allowing for ½" excess to form a loop. With round-nosed jewelry pliers, make a loop that tucks back into the uppermost bead or bamboo after connecting it to the looped-wire base. Close the loop of the bamboo hanger with flat-nosed pliers.

Step Six: Ear wires connect to the upper loop-simply open the ear wire loop, string it through the loop-wire base and close with flat-nosed pliers.

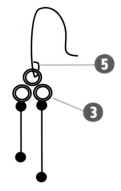

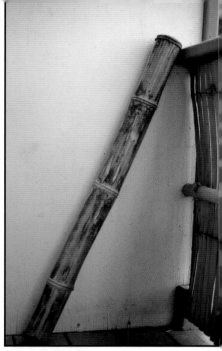

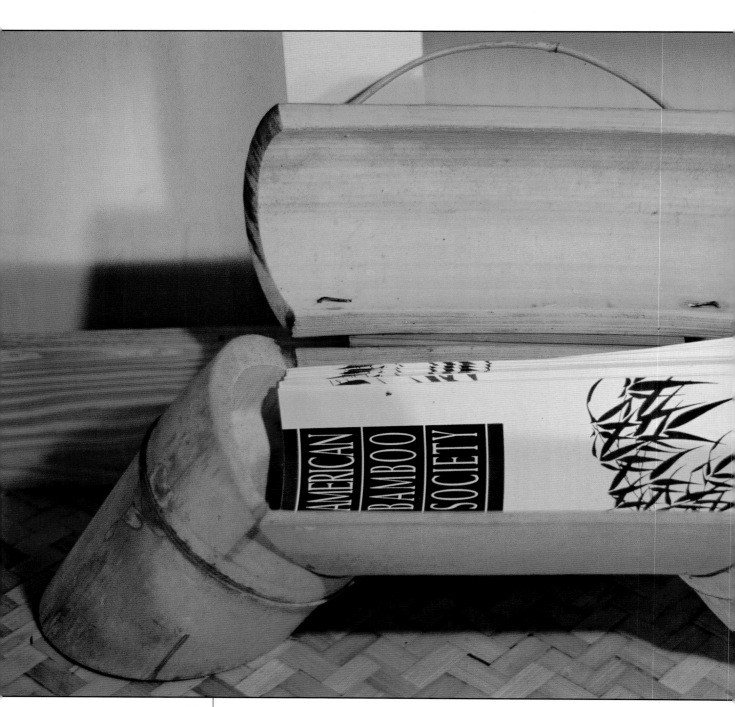

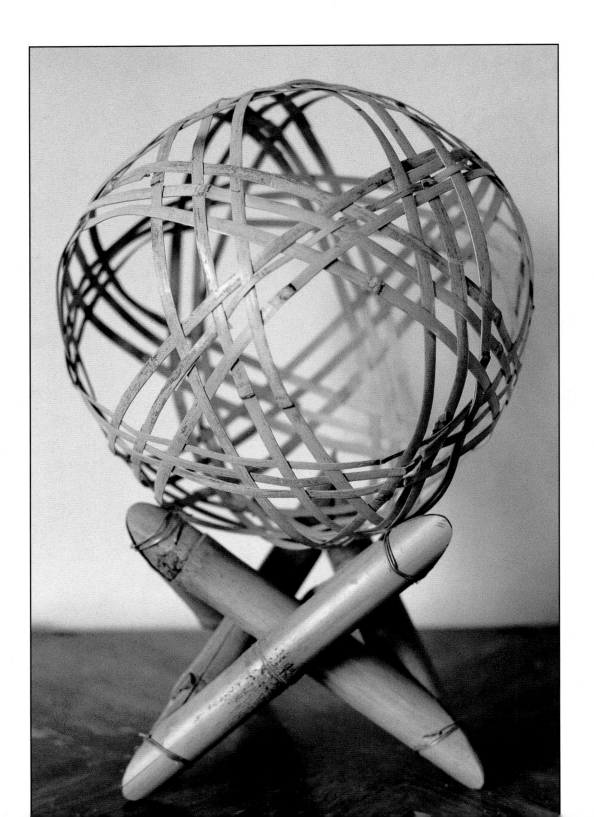

CLOCK

EASY

Bring some bamboo character to the standard clock face with this design that is perfect for adding a bit of rustic flare to your home.

Materials

- ∿ 18"-long, 2"-diameter bamboo
- ∿ 2 1/2"-long, 1"-diameter pieces of bamboo
- ∿ 16"-long, 1"-diameter bamboo
- ∿ 1 quartz, battery-operated clock mechanism
- ∿ 18- to 20-gauge galvanized or brass wire (10')

Tools

- ∿ Knife
- ∿ Vise Grips
- ∿ Drill and Drill Bit Set
- ∿ Wire Cutter

∾ Figure 1 ∾

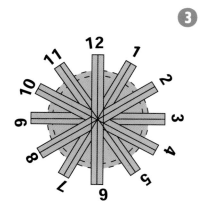

∾ Figure 2 ∾

Step One: Split all the bamboo pieces in half. (See page 16 for splitting techniques.)

Step Two: Arrange the bamboo halves as shown in figure 1. Cut out the center sections of the bamboo pieces so they can be stacked tightly—the 11 o'clock and the 1 o'clock piece first, the 10 o'clock and the 2 o'clock second, then the 3 o'clock piece and finally the 12 o'clock piece.

Step Three: Drill a hole through the center of the 12 o'clock bamboo piece and thread a wire around the remaining bamboo pieces so they are held tight and have a static configuration of a clock face (fig. 2). (There are many ways of doing this, and, since the wire is hidden behind the 12 o'clock bamboo, the tangled web will not be seen.)

Step Four: Attach the clock mechanism to the center of the 12 o'clock piece. Set the clock and hang it. Now you have the time!

HANGING PLANT HOLDER
EASY

Bamboo yields itself to easy splitting and is a natural for displaying plants—indoor and outdoor. This project can be accomplished in a short time and requires few tools.

Materials

- *6'-long, 3"-diameter, thin-walled bamboo with at least two nodes*
- *18- to 20-gauge galvanized or brass wire (2')*
- *6'-long clothesline or nylon cord*
- *1 planting pot (12" diameter)*

Tools

- *Pruning Saw or Mitre Saw*
- *Knife or Machete*
- *Vise Grips*
- *Wire Cutter*
- *Drill with 1/4" Bit*
- *Metal Pipe or Broom Handle*

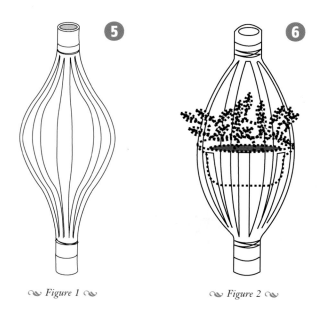

Figure 1　　　　　*Figure 2*

Step One: Cut the ends of your 6' bamboo just below a lower node and just above an upper node.

Step Two: With a mop handle, metal pipe or other handy instrument, knock out the nodes through the complete length of your bamboo. This eases the process of splitting and prevents the accumulation of water at the ends.

Step Three: Wrap two to three loops of wire around the bamboo at either end, just inside the node. This will prevent the splits from continuing through the end and wrecking the integrity of the piece.

Step Four: With a knife, initiate a cut at the center of an internode and just to the side of the branch scar (we are trying to avoid making a split directly through a branch scar). Once started, the cut will split in both directions. By twisting the knife, the split can be worked through the nodes to either end.

Step Five: Repeat the process on the opposite side of the culm. Again, initiate a split so the culm is split into quarters, then eighths (fig. 1).

Step Six: The bamboo culm, now split into eight sections, can be pried apart to hold a small plant pot placed in the center. Arrange the eight splits equally around the plant pot. It can be slipped down so that a node supports the pot. If the bamboo needs to be strengthened around the pot, the lower wraps of wire can be slid up, making the bamboo tighten around the pot (fig. 2).

Step Seven: Drill a $1/4$"-diameter hole just above the top node, to the opposite side. Run a loop of cord through the holes and tie a loop. The hanger can now be hung from your favorite spot. (The knot can be hidden inside the bamboo culm by inserting the cord into either side and tying inside.)

52

FURNITURE & ACCESSORIES

LAMP SHADE
ADVANCED

The straight grain of bamboo is one of its most distinctive qualities and it is highlighted in many traditional Asian designs. This bamboo lamp shade splits bamboo to a fine degree.

Step One: Select *Bambusa vulgaris* or *B. tuldoides tulda* bamboo because of the species' straight grain, which makes them easy to split. With a mitre saw, cut them to fit the specified measurements (fig. 1).

Step Two: Wrap two to three loops of brass wire just above the lower node of the 12" bamboo. Tighten, then twist the end, and press against the bamboo. This will prevent subsequent splitting from continuing through the node and damaging the piece (fig. 1).

Step Three: With a sharp knife, make an incision at the upper end of the bamboo, pressing the knife to the depth of the blade and then twist. The split will continue to just above the wire loop. Repeat this procedure on the opposite side of the bamboo culm.

Step Four: Proceed with this technique, splitting the culm into $1/4$s, $1/8$s, $1/16$s, $1/32$s and finally $1/64$s. With each step, you work your way around the culm, splitting each subsequent section in half. With each round, the number of splits doubles—first two, then four, then eight, then sixteen, and finally thirty-two. It is important to be accurate in splitting; when finally at sixty-four splits, each individual split is fairly similar to all others. It helps to press a bamboo ring of similar diameter into the center of your culm, splaying out the splits and easing the splitting process (fig. 2).

Materials

- *12"-long, 2"-diameter, $^3/_{16}$"-walled* Bambusa vulgaris *or* B. tuldoides tulda *with a node on each end*
- *4"-long, $2^1/_4$"-diameter bamboo*
- *18- to 20-gauge brass wire (12")*

Tools

- *Mitre Saw*
- *Knife*
- *Wire Cutter*
- *Flame*

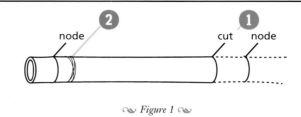

node   node   cut   node

Figure 1

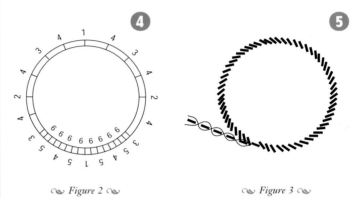

Figure 2   *Figure 3*

Step Five: At this point, each split should be slightly less than $^1/_{16}$" in width. Pressing your bamboo ring as far as practical will splay the split even more and facilitate the next step.

Step Six: Cut a length of 18- to 20-gauge brass wire that is more than twice the circumference of the splayed bamboo. Bend it in the center to make a double-stranded wire. Starting anywhere on your splits, about $^1/_2$" from the end, twist the wire around the first split. Grab the next split and twist the wire again around it in the same manner, keeping the wire tight and the space between the splits at a minimum. The idea here is that the splits will line up, thin edge to thin edge, which is a 90-degree twist from where they originate in the culm. The effect is a whorled and splayed shape. Also, the final diameter of the shade is dependent on the width of the wall. The bamboo chosen determines the final shape (fig. 3 and see photo).

Step Seven: Continue twisting the wire around each subsequent split, unraveling the remaining wire periodically. When the circle is complete, continue through the initial twisted wire a couple of splits, as an overlap. Twist the wire, cut off the excess, and tuck the twist against the splits.

Step Eight: Stretch and tighten the wire by pushing out the bamboo splits with your hands and shape to a circle. The bamboo ring that is inside the culm can now be removed.

Notes on Splitting: Very large split
and splayed bamboo lamp shades can be achieved using slightly larger-diameter bamboo, and splitting to 128 separate splits. This is obviously more difficult and requires fine, straight-grained bamboo. (See page 16 for splitting technique.)

Splitting through a node can be accomplished, but not to the degree of sixty-four splits. The grain can be erratic through a node, with splits often zigzagging. I would not go more than sixteen splits through a node, depending on the bamboo. Additional splits can be achieved above the node but only to the degree the bamboo will allow. However, when approaching very thin splits, if the initial cut is not perfectly centered, the split may wander to one side or the other and break off. This can be tolerated to a degree without a noticeable degradation of appearance.

Sometimes, depending on the diameter of bamboo, the sixty-four splits are still very stiff and resistant to twisting. This makes the wire looping process somewhat difficult. If you go around and twist each split by hand to 180 to 270 degrees, the split is stretched a bit and springs back to a twist much closer to the 90-degree twist that you need.

In the splitting process, the bamboo will exhibit tiny hairs of split-off grain along the surface of the splits. These can be made to disappear by quickly passing the bamboo over a flame and burning the hairs away. The heating process also relieves some of the stress of the splitting and splaying process, relaxing the bamboo and making it more comfortable with its new shape. However, do not dawdle—always keep the bamboo moving through the flame to prevent darkening and possible ignition.

STAND ADVANCED

A tripod is a simple and elegant way to cradle potted plants or display dried flowers. It also can serve as a base for a lamp.

Materials

- 12"-long, ¹/₄"-walled bamboo
- 2 bamboo branches, ¹/₄" to ¹/₂" diameter and twice the length of your base.
- 18- to 20-gauge brass wire (24" to 36")

Tools

- Mitre Saw
- Metal Pipe or Similar Object
- Wire Cutter
- Pencil
- Knife
- Vice Grips

Step One: With a mitre saw, cut a 90⁰ cut about 2" below the node of your bamboo (fig. 1).

Step Two: Knock out the membrane at the nodes with a long piece of rebar (metal pipe) or another similar object.

Step Three: A short stand may only require a splay through one internode. A taller application may require splits to splay through two internodes. Depending on your project, wrap two to three loops of wire around the bamboo culm just below the node where you want the splayed splits to stop (fig. 1). To secure the ends, twist and tuck under.

Step Four: With a pencil, mark the base of the culm such that it is split into thirds. To prevent the possibility of splitting through a branch scar later, make sure one of the marks is adjacent to the branch scar.

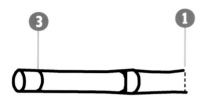

Figure 1

Figure 2

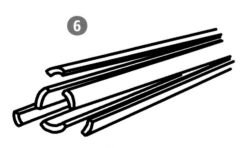

Figure 3

Step Five: Make the splits using a knife worked into the culm to the depth of the blade and twisting to cause the split to continue up to the point where the wire binds the culm (fig. 2).

Step Six: The next cut is a judgment call. You will need to determine the width of the splayed legs of your tripod base. This is dependent upon the thickness of the bamboo culm (thicker means stiffer, and a narrower split would help alleviate the stiffness). Also, whether the split travels through one or two internodes will have an effect on your chosen leg width. (A two-internode split is easier to splay out.) Make your call, and cut the next splits to one side of the previous three splits—anywhere from $5/16$" to $3/4$" (fig. 3). You will now have the culm split into six sections (three legs and three central sections that are not splayed).

Step Seven: Splay one leg out to where it is sufficient to support your project but not with such a stressful bend that it will break. Measure the distance from the inside of the leg just above the node to the opposite culm split, which is part of the central shaft (fig. 4).

Step Eight: Cut three bamboo branch segments of the measured length. One end is cut at a 90-degree angle, the other at approximately 15 to 20 degrees from the 90-degree angle, or equal to the angle that the splayed leg makes (fig. 6).

Step Nine: Insert the branch sections between the splayed legs and the central shaft with the angled end rotated to the angle of the leg and seated just above the internal node ridge (fig. 5).

Step Ten: Grasp the three internal shaft sections and draw them together to tightly secure the branch sections radiating out to the legs. Wrap two to three loops of brass wire around the shaft just below the bottom branch (they are stacked one on top of the other) and tighten. Twist the wire with vice grips and cut off the excess, tucking in the twisted wire.

Step Eleven: At the splayed leg-branch intersection, drill a hole through the splayed bamboo such that it emerges in the center of the branch. Fit a small bamboo dowel into this hole to hold the branch in place. Do this for all three legs.

Step Twelve: The splayed legs will now be shorter than the central shaft. Therefore, the central shaft must be cut to allow the stand to be supported by its legs. Take a straight edge, place it against the bottom of two legs, and line up the third behind with your eye. This will tell you the minimum amount of cut you need to make on the central shaft. If the bamboo branches are in the way, slide them up, one at a time. Cut the excess bamboo from the central shaft, slightly more than the minimum required.

Step Thirteen: Now stand your stand. If it is not perfectly vertical (look at it from all angles), shorten the appropriate leg slightly (not too much). The central shaft should still clear the floor; otherwise you may need to shorten that as well. The taller the project (floor lamp vs. table lamp), the more the legs need to be splayed.

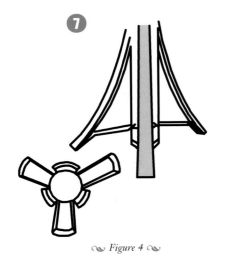

❼

~ *Figure 4* ~

❾

~ *Figure 5* ~

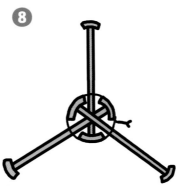

❽

~ *Figure 6* ~

59

READING LAMP

The utility of this design makes this lamp a practical addition to any interior. The design adheres to the philosophy of showcasing the building material's best qualities by making use of the straight, splitable grain, its hollow, tubelike form and bamboo's inherent beauty. This project combines building a lamp shade (page 54) and a stand (page 57) for the total bamboo experience.

Materials

- 2, 12"-long, 2"-diameter, $^3/_{16}$"-walled Bambusa vulgaris or B. tuldoides tulda *with a node on each end*

- 2 bamboo branches $^1/_4$" to $^1/_2$" diameter and twice the length of your base.

- 4"-long, 2$^1/_4$"-diameter bamboo

- 18- to 20-gauge brass wire (8')

Tools

- *Mitre Saw*

- *Metal Pipe or Similar Object (Step Two)*

- *Wire Cutter*

- *Pencil*

- *Knife*

Step One: Cut off a single internode length from just below a node to just below the next node. This will be the swiveling lamp shade (fig. 1). Follow the procedure on pages 54–56 for the split bamboo lamp shade.

Step Two: Make the bamboo stand on pages 57–59.

Step Three: With a hole saw of appropriate diameter, drill a hole through the upper internode of the tripod base approximately 2" above the node. The diameter of the hole should be such that half the circumference of the bamboo culm is cut through. Remember to remove the hole cut out of the front side and to repeat the procedure from the opposite side using the pilot hole drilled originally (fig. 2).

Step Four: Using a sharp knife, cut out the culm splits above the hole saw cuts. Sand and round off the top of the two protruding tongues of the culm. Also, round off the edges. Bamboo can have a very sharp edge, sharp enough to cut (fig. 3).

Step Five: Cut a ring of 2" width from a larger culm so it can be slipped over the lamp shade and act as a support for an adjustable lamp. Wrap the ring with two loops of wire on either side of the center. Drill a $^1/_4$" hole on opposite sides of the ring (or an appropriate-sized hole to accept a tight fit with

Materials cont.

- 16"-long, 2"- to 3"-diameter, ¼"-thick-walled bamboo with three internodes

- 6'-long, 4"-diameter bamboo

- 18- to 20-gauge brass or galvanized wire (10' to 12')

- 3 bamboo branch sections ¼"- to ³⁄₈"-diameter. (Three for the base, plus dowels for swivel.)

- Electrical components: pigtail socket, toggle switch, wall plug, 10'- to 12'-lamp wire and electricians tape.

- Glue

- Sandpaper

Tools

- Mitre Saw
- Hole Saw Kit
- Vise Grips
- Knife
- Wire Cutter
- Drill and Drill Bit Set
- Flame

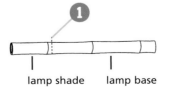

lamp shade   lamp base

Figure 1

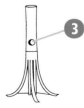

Figure 2

Figure 3

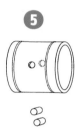

Figure 4

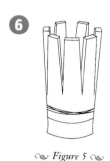

Figure 5

a handy bamboo branch), to be used as a dowel swivel. Cut two branch dowels about ³/₄" in length (fig. 4).

Step Six: For the lamp socket holder you have two choices depending on the size bamboo you have available. If the bamboo is complementary to the shade and will slide nicely into the split shade, do the following: cut just below a node and approximately 4" above the node. Wrap two to three loops of wire around the bamboo just above the node. Twist, tighten and cut off the excess wire. The upper section can be split into eight sections so that it will hold the lamp socket tightly within its splayed splits (fig. 5). If your bamboo is complementary to the socket (it fits nicely), but is too small of a diameter to fit tightly into the lamp shade, the opposite end can be splayed out. Split into eight sections and splay out with a similar diameter ring. This will be flexible and hold tightly inside the shade. (See page 16 for splitting techniques.)

Step Seven: Knock out the node on the lamp socket holder piece so that a wire can be run through it. Push the lamp holder piece down into the split bamboo shade of the lamp. It should fit tightly, pushing out the split bamboo, and it should be of a final length such that the bulb does not protrude beyond the end of the shade. (I highly recommend a 40-watt, clear appliance bulb. It imparts a nice shadow effect through the split bamboo, it is not too hot and it is a smaller size.)

Step Eight: Cut a length of lamp wire of approximately 6' to 8', and connect to the pigtail plug according to accepted wiring techniques. (I twist the wire together and wrap with electricians tape.)

Step Nine: Fish the end of the wire through the socket holder, down through the lamp, down to the base, and let it emerge between two of the branch bamboos used to splay the tripod

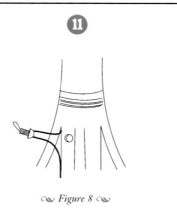

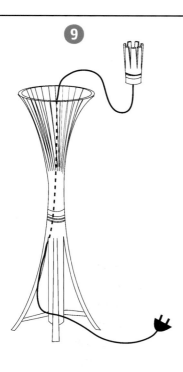

~ Figure 8 ~

~ Figure 9 ~

base (fig. 6). Add a plug to the end of the wire, and push the socket holder snugly into the split bamboo shade. (You may glue the interface of the socket holder and the lamp shade.) Center the socket holder.

Step Ten: Drill a similar-sized hole on the lamp shade ring, centered on the end of each of the protruding culm tongues (fig. 7).

Step Eleven: Assemble the components as shown in (fig. 8). Adjustments may be made to get a configuration that will work. Ideally, the lamp shade can be pointed directly vertically, sandwiched between the two protruding culm tongues. The end of the shade may need to be tapered or the tongues may need to be shaved wider. Glue may be needed to augment the friction fits of this design.

Step Twelve: Drill a hole to accept the toggle switch centered between the node and the hole saw cut, and install toggle switch wired in series (fig. 9).

Alternatives: A floor version of this lamp is very elegant and practical. Care must be taken such that the torchère configuration is approximately 6'. If you do so, you may want to cut a longer measurement of lamp wire for step 8. Also, a two internode splayed tripod base is recommended for more stability.

~ Figure 6 ~

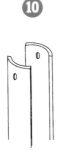

~ Figure 7 ~

SCREEN DOOR ADVANCED

Lightweight, functional and more original than store-bought designs, this screen door sets the tone for coming home.

Materials

- 2 straight-grained, thin-walled medium diameter (2' 2 ½") bamboo culms approximately 85" long

- 2 straight-grained, thin-walled medium diameter (2' 2 ½") bamboo culms approximately 45" long

- Black, nylon screening (36"-wide, 45"-long)

- 2 wood strips to fit inside the bamboo components, approximately 12' long

- Wire (18-20 gauge, brass, or galvanized)

- Wood/deck screws

- ³/₈" staples

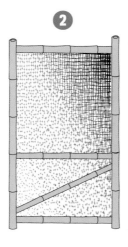

∽ *Figure 1* ∽

Step One: If your door opening is set, design a door with outside dimensions allowing ½" to 1" clearance all around. If the opening is square, plumb and straight, ½" is fine. If, however, the door opening is of bamboo, it is no longer straight, and a 1" clearance may be more practical.

Step Two: Collect complementary bamboo with the lengths needed for your door, according to the illustration in figure 1.

Step Three: Cut the vertical bamboo to length. Decide how the bamboo will be best oriented. Using a hole saw, cut the proper sized holes to accept the horizontal members. The horizontal bamboo will need to be fit into its hole on the verticals to determine the length needed to achieve the proper width. How deeply the horizontal bamboo seats determines the length. Cut and re-cut the three horizontal members.

Step Four: Assemble the door on the floor, making sure the horizontals are seated tightly. Rotate the horizontal members so that

Tools

- Table Saw
- Mitre Box/Saw
- Drill
- Hole Saw Kit
- Drill Bits
- Staple Gun
- Hammer
- Vise Grips
- Wire Cutter
- Countersink Bit

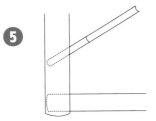

Figure 2

Figure 3

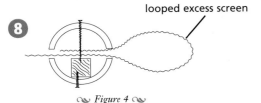

looped excess screen

Figure 4

they are aligned flat or at least the best orientation. Square up the layout by measuring diagonals and getting them identical.

Step Five: Using a thinner, but stiff and straight piece of bamboo, lay out a diagonal between the vertical bamboos on the lower panel 2" to 3" away from the joint between the horizontal and vertical members. Lay this out such that the lower end is on the hinge side of the door. It will prevent the door from sagging and be under compression. Mark this accurately, and hole saw a hole to receive this diagonal in the respective vertical bamboo culms (fig. 2). Cut the diagonal by fitting into the holes and butting the outer wall of the verticals.

Step Six: With the complete frame tightly laid out on the floor and all components rotated in their final orientation, start splitting the bamboo at the upper corner, making sure the splits are matched in the inside corner (fig. 3). The outer split is not as important but should be matched and the culms equally split into halves.

Step Seven: Work your way around, matching the previous inside split to the next split. After splitting three sides of the top opening, the fourth split on the horizontal is done. If you are very lucky, the splits will meet at the fourth corner perfectly. More likely, they will be off slightly. Rotate that horizontal to split the difference, then continue in the same manner through the rest of the door. Splits will not continue through the holes you have drilled for the horizontals, so that they need to be re-started. That is good so that you can again match up splits. The diagonal is also split, matching it with the splits on the verticals.

Step Eight: Open up the splits, and knock out the node septums to allow a strip of wood, about $1/2$" by $5/8$", to be inserted inside the culm. This wood will serve as an anchor to staple the screen, and an anchor to re-secure the bamboo halves together.

Step Nine: With a table saw, cut strips to fit into the fullest length of each door component. Cut to length, and check to see if they do indeed fit with no overlaps and no gaps.

Step Ten: Layout the door with the upper split half off and the wood strips inserted. Lay your screen (black nylon is best, it stretches), centered on the door. Keeping the screening taut, staple it to the wood strips with $^3/_8$" staples. Use a hammer if necessary to make them tight. As an easy way to weather strip the gap between the door, the floor and the jambs, I loop the excess screen back into the bamboo, leaving enough exposed loop to span the gap (fig. 4).

Step Eleven: Progressively, join the bamboo with the top half. These can be secured with wire loops carefully threaded through the screen, or screwed. The bamboo is pre-drilled, countersunk, and screwed into the wood from both sides. The ends of the verticals should be wired, as they need to be extra strong.

Step Twelve: I have used burlap for the lower panel, which looks great, and stretches nicely.

Step Thirteen: With the assembly completed, fit the door into its opening. Center it. The best hinges are those suggested for the bamboo gate, a rotating fulcrum mounted on the floor, centered

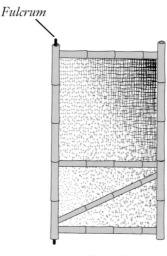

Fulcrum

∿ *Figure 5* ∿

on the vertical door bamboo. A guide fulcrum is needed at the top as well. A bamboo hinge can be done with thin bamboo shafts. Any easily installed arrangement will work.

Step Fourteen: A bent bamboo handle is needed, and you could put in a bamboo door latch, to keep the door closed.

Alternatives: Bamboo framed screen panels can be created using the same strategy. I have split the bamboo off- center, however, as this facilitates insertion of the wood strip, the screen, and easy replacement of the cap. Other panels can be used as well, even somewhat rigid and thicker panels. Some extra strip may need to be removed to make these fit nicely. But, as these panels are dimensionally stable, there is no need for diagonal bracing (fig. 5).

BED FRAME
ADVANCED

Materials

- Select the largest diameter available and complementary bamboo for your design

- 4-80" posts

- 4-80" complementary horizontals

- 2-96" complementary horizontals

- 2-pieces at 80", and 8 smaller diameter pieces up to 40" for the headboard (fig. 1)

- Sufficient bamboo for the split cross supports (no box spring)

- One 4' by 8' by ³/₄" sheet of plywood, and two 8' by 2" by 4" boards to build a mattress frame support

- Woven grass or bamboo matting sufficient to cover frame

- Contact cement

- 8-6" carriage bolts

- Four lengths of thin bamboo for post dowels (2" by 2" by 4" by 8')

- Glue

- Deck screws (1 ⁵/₈")

The ultimate piece of furniture for total immersion into the bamboo world is the bamboo bed. What better way to sleep and dream of bamboo projects? The designs for a bamboo bed are of infinite variety. However, certain design criteria are universal—it should be stable, strong and aesthetically pleasing. The king-size bed presented here follows the example of West Indian furniture designers and is high off the floor.

Step One: Assemble the frame components as shown in figure 2. (You could paint or varnish the plywood and cap the end grain with bamboo.)

BED FRAME CONT.
ADVANCED

Tools

- *Table Saw*
- *Power Mitre Box*
- *Router*
- *Drill*
- *Drill Bit Set*
- *Hole Saw Kit*
- *Knife*
- *Measuring Tape*
- *Round over Router Bits ($^3/_8$", $^1/_2$")*
- *Spade Bit Set*

Step Two: Cut the two 8' by 2" by 4" boards in half lengthwise, to make 4 pieces at 8' by $1^1/_2$" by $1^5/_8$". Cut the plywood into four 10"-wide boards. The $1^1/_2$" by $1^5/_8$" pieces are glued and screwed flush to the bottom edge of the plywood with end details as shown in figure 3. Cut the cut off ends of the $1^1/_2$" by $1^5/_8$" to $8^1/_2$" long and rip at a 45-degree angle to get the corner wedge pieces. Note that before making these cuts, you need to measure your box spring to ensure it will fit in the box you are building.

Step Three: Pre-drill the corner screw connections, and screw the four sides together as shown in figure 3. The sides will need to be finished before final assembly. Rout the top edge with a $^3/_8$" round over, both sides, and use a $^1/_2$" round over bit to round over the lower, outer edge as shown in figure 2.

Step Four: A veneer matting is wrapped easily around the curved edges, using contact cement. Excess matting can easily be cut off along the inside edge of the $1^1/_2$" by $1^5/_8$" board (fig. 5). The very ends of the frame need not be covered, as they will be capped by the bamboo posts.

Step Five: Select your frame height off the floor. The frame should be centered on the internodes of the four bamboo posts. Mark the $10^1/_4$"-frame space (allowing some play for easy assembly), measure down and mark where the post ends, and measure up to where the top of the post will be. Cut the ends accordingly. The frame cutout on the post should be cut deep enough for the frame to be supported, but not so deep that the post will interfere with the mattress. Cut the top and bottom slits, being careful to end the cuts parallel to each other. Split out the piece with a knife (fig. 4).

Step Six: Fit the post to the frame and adjust if necessary. The post is secured to the frame with carriage bolts, tightened from the inside. Mark where the carriage bolt holes will be, and drill a countersink with a proper sized spade bit. This will recess

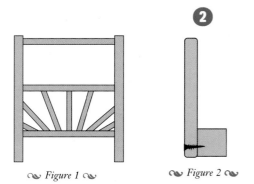

②

Figure 1 *Figure 2*

the nut into the wood. Drill holes for two bolts at a 45-degree angle, centered on the corner, according to figure 6.

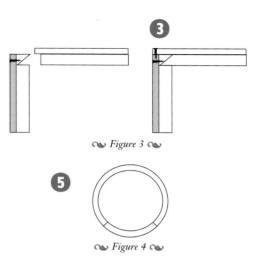

Figure 3

Step Seven: A spacer may be needed between the frame and the inside wall of the bamboo. This will be a custom fit and size, depending on the diameter of the bamboo, wall thickness, and depth of the frame cutout. A square, centered on the bamboo cutout, will give a fairly accurate measure of the distance between the wall and the frame corner. The spacer can be made on the table saw to fit (fig. 7). The spacer is drilled through to match the frame holes for the carriage bolts. Finally, with all parts in place, the carriage bolt hole can be

Figure 4

drilled through the outer wall of the bamboo. This is done slowly to prevent splintering. A pilot hole can be done first, with the final diameter drilled from the outside. Another alternative is to lightly tap the carriage bolt as it rests against the inside bamboo wall and passes through the frame holes. Remove the bamboo, and the indentation on the inner wall will mark where the carriage bolt hole is drilled.

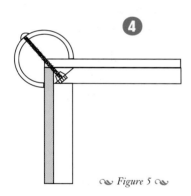

Figure 5

Step Eight: Assemble the frame and four posts, loosely securing with the carriage bolts. The upper horizontal, complementary bamboo pieces are fitted next. The side pieces are higher than the end pieces with approximately a 2" space between the pieces where they connect to the post.

Step Nine: Measure the distance between the posts at the frame, when tightly fitted. This distance should be duplicated at the top. The horizontal bamboo is cut such that there is a clear internode end on either end to facilitate insertion into the posts. With a matched diameter hole saw, cut holes into the posts. Adjust the length of the horizontals such that, when tightly butted inside the posts, the distance between the posts duplicate the distance between the posts at the frame. Repeat this procedure for each of the four horizontal bamboo pieces.

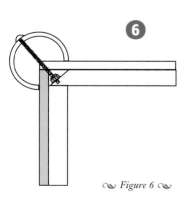

Figure 6

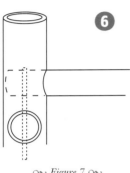

✑ Figure 7 ✑

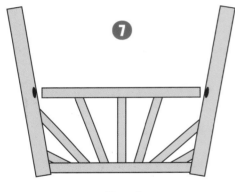

✑ Figure 8 ✑

Step Ten: The securing of the horizontal bamboo pieces inside the posts is done with a vertical branch or thin bamboo dowel. Cut the dowel long enough to cover the distance from the top of the side horizontal, to the bottom of the end horizontal. It should also be cut such that the node is on one end, with a long internode for easy insertion. With the horizontal bamboo pieces rotated to the desired final orientation, carefully drill a hole centered on the horizontal and the vertical post. Continue through both horizontal bamboos, keeping the holes aligned. The dowel can then be dropped down through the holes, and lightly tapped into place. This is removable, and is a very easy way to secure the three-way joint.

Step Eleven: The headboard is placed according to your design. Allow sufficient space between holes such that you will minimize the tendency to split (3" to 4" minimum). The lower horizontal should be located at about the same height as the upper surface of the mattress. Maximize the component lengths such that they can be fitted, but will not pop out when in the final placement. Verticals intersecting with the upper horizontal can be cut to the max, all then inserted into the top horizontal. The horizontal is then inserted into the remaining post. By not inserting the dowels securing the upper horizontals, the posts can be spread apart to facilitate the insertion of the headboard.

Step Twelve: When all components are fitted and situated, the dowels are tapped into place, and the carriage bolts finally tightened. Any unforeseen fitting problems will be seen now, and can be attended. If wire reinforcement is deemed a good idea, do it now. The tops and bottoms of the posts are likely spots, as well as around the frame inserts.

Step Thirteen: The box mattress and the regular mattress can be dropped into place. Ready for a rest?

Materials

- Enough larger-diameter, thick-walled bamboo at least 72" long, and that spans 48" when placed side by side

- Three pairs of matching bamboo culms, successively complementary to each other. The largest at 24", next at 12" and the smallest at about 12"

- Two lengths of ⅛" galvanized cable, about 12'

- 8-⅛" cable clamps

- 4 lengths of thin bamboo (cable sheathing—optional)

- 2 ceiling hooks (according to ceiling structural parameters)

- 2 S-hooks

Tools

- Saw
- Hole saw set
- Drills
- Wrench
- Drill bit set
- Wire cutter

This porch swing is designed to conform to your body as you rock gently in the late afternoon sun.

Step One: Select enough larger-diameter bamboo of about 72" in length to span a width of 48" when placed side by side.

Step Two: Cut the bamboo at a shorter length, around 60", centering the cuts equidistant from the end nodes. It is best to have nodes close to the ends of the bamboo length for strength.

Step Three: Arrange the bamboo culms such that nodes are offset from adjacent culms, and individual culms are matched in character (similar size, color, bend). The idea is to have the bamboo culms conform to each other, with no large gaps due to opposing curvatures. Each culm should be rotated to best match its neighbors (fig. 1).

Step Four: With a pencil, and the culms lined up and oriented to your satisfaction, make a mark about 4" in from the ends of the culm, where a hole will be drilled through the culm to snake a galvanized wire. The holes should be perfectly lined up, and bisect the culm diameter. The front, last bamboo is drilled with only one hole (fig. 1).

Step Five: For the arm rests, select and cut three pairs of matching culm sections, according to the diagram in figure 2. The three sections should be complementary to each other, and fit into the corresponding bamboo piece as shown. Lengths of each is determined by fitting to the conformation desired. Generally, though, the front arm support should be about 12" in height, allowing for insertion depths. The horizontal arm

Figure 1

should be approximately 24". The triangulating support is cut such that the angle between the support and the arm is maintained at 90 degrees (fig. 2).

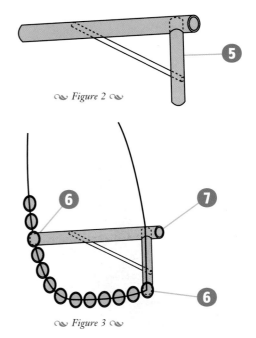

Figure 2

Step Six: With a hole saw, cut the holes in the bamboo sling to receive the arm support arrangement. A decision must be made as to which bamboo on the backrest portion of the sling is to receive the armrest. This decision determines how much the curvature the seat has. This is a judgment call, and is not written in stone. The final criteria is when you sit in the swing. If it needs to be adjusted, a bamboo support or two can be added or removed (fig. 3).

Step Seven: Drill a hole in the center of the armrest, where the cable will emerge from the armrest support. Also, a cable hole through the ends of the supports may be needed for the cable to pass through (fig. 3).

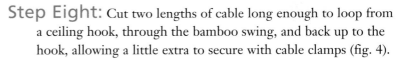

Figure 3

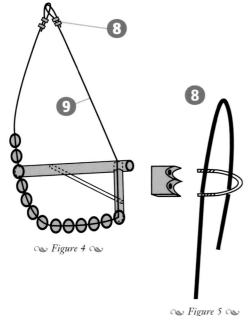

Figure 4

Figure 5

Step Eight: Cut two lengths of cable long enough to loop from a ceiling hook, through the bamboo swing, and back up to the hook, allowing a little extra to secure with cable clamps (fig. 4).

Step Nine: Snake the cable through the adjacent bamboo culms of the sling, and up through the arm support (fig. 4).

Step Ten: With the swing supported from below at the proper height for a swing, (your call), secure the cables with clamps, and suspend the swing (fig. 4).

Adjustments will be necessary! The swing can be slid one way or the other along the cables to get the proper angle for a seat. The loops of cable may need to be lengthened or shortened to get a comfortable swing height. Tighten the clamps in their final position. "Never saddle a dead horse." This is a longshoremen saying describing the proper method of securing a cable clamp. The saddle, or piece that is placed over the horseshoe metal piece, is always placed on top of the cable that is supporting the weight of the longer one (fig. 5).

72

OUTSIDE
& GARDEN

WATERSPOUT

Materials

- 3'-long, ³/₄"-internal-diameter bamboo
- 6"-long, 1¹/₂"-diameter bamboo with a node at one end
- 18- to 20-gauge galvanized or brass wire (3')
- 5'-long, ¹/₄"-wide plastic tubing
- Medium-sized water pump

Tools

- Mitre Saw
- Hole Saw
- Drill and Drill Bit Set
- Vise Grips
- Wire Cutter
- Knife

The waterspout is necessary for any bamboo water accessory. It provides the soft, soothing rhythm of cascading water and a steady stream for your bamboo water wheel.

Step One: Split the vertical 3'-long bamboo piece in half. Whittle any node obstruction from the central cavity such that the plastic tubing can pass through the bamboo.

Step Two: Re-assemble the vertical bamboo with the plastic tubing inside and protruding from the top about 3" and the bottom such that it will connect to your water pump easily. Secure with wraps of wire at the top and bottom (fig. 1).

Step Three: With the mitre saw, cut a diagonal slice from the horizontal bamboo piece on the opposite end of the node for the water to flow out (fig. 2).

Step Four: Cut a hole to receive the vertical bamboo just inside the node end and on the opposite side of the spout mouth.

Step Five: Snake the plastic tubing into the horizontal piece toward the mouth, and then insert the vertical piece. The tubing should not protrude out the mouth of the spout (fig. 3).

Step Six: Place the spout in the water, secured by rocks and connect to the hidden water pump.

(If a trickle of water dribbles down from the hole in the horizontal bamboo, you can neatly wrap some electricians tape around the vertical bamboo at the site and make it watertight.)

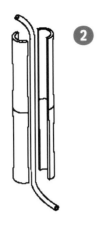

Figure 1

Figure 2

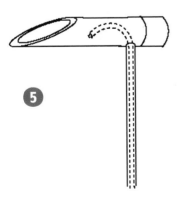

Figure 3

Materials

- 1 large-diameter culm with 18 internodes (16 for cups, 1 for central axle, 1 for radiating cup supports)
- 18- to 20-gauge galvanized wire
- Heavy-gauge wire (clothes hanger) for central axle
- Axle support material (bamboo or site-specific supports)
- Elastomeric roof coating
- Water-based polyurethane
- Bamboo trough with recirculating pump and pond
- Galvanized round head wood screws (these don't split bamboo)
- Trough

Tools

- Mitre Saw
- Drill and Drill Bit Set
- Paintbrush
- Vise Grips
- Wire Cutter
- Knife

Some say that bamboo water wheels and troughs were instrumental in not only the irrigation of rice fields but also the agricultural revolution. The hollow, cylindrical nature of bamboo is a natural container for water. If the containers are arranged in a circle, surrounding an axle, they can catch water as it flows and convert the energy to do useful work. The conversion of water flow to mechanical motion is in itself a beautiful thing.

Step One: Depending on the size of the wheel to be assembled, select a bamboo culm with a diameter adequate to catch the water flow from your trough.

Step Two: Using a power mitre saw, cut cups from each culm node/internode interval. Cut at 90 degrees the cup height at about the same distance above the node (the bottom of the cup) as the diameter of the culm. Cut the culm at 75 degrees, slightly below the node. The angle is sufficient to tip the cup out from the wheel, and thus catch more water (fig. 1).

Step Three: The central axle will be another culm of similar diameter as the outside cups. A convenient length and cut is just above and just below two consecutive nodes to get a full internode section.

Step Four: Wrap 2 to 3 loops of galvanized wire just inside one of the nodes. Twist and tighten with vise grips, and cut off the excess wire.

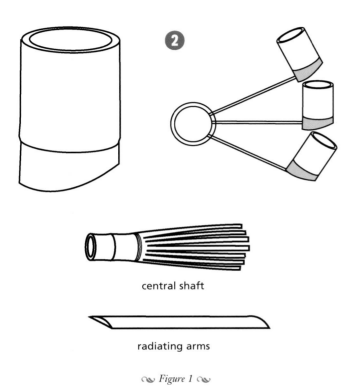

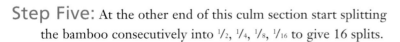

central shaft

radiating arms

❧ *Figure 1* ❧

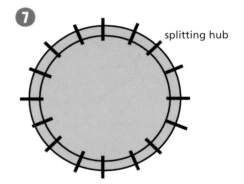

splitting hub

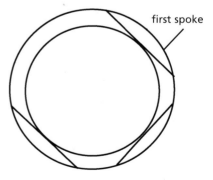

first spoke

Step Five: At the other end of this culm section start splitting the bamboo consecutively into $1/2$, $1/4$, $1/8$, $1/16$ to give 16 splits.

Step Six: Using another culm internode section of similar diameter (no nodes), split it into 16 sections as well. Flatten these splits with a knife.

Step Seven: Insert each flat bamboo split into the whole culm axle splits, taking care to not insert them too far into the center (allow space for all the inserts). The expanded end of the axle (without a binding wire) is now gathered together with 2 to 3 loops of galvanized wire. Bringing the end back together tightly, securing the radiating splits in their position (fig. 2).

❧ *Figure 2* ❧

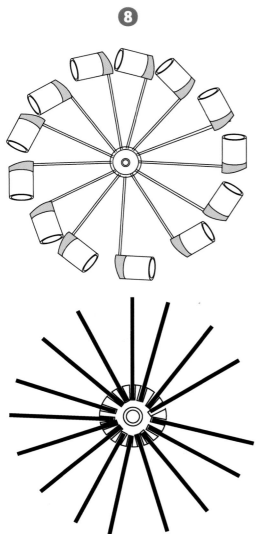

Step Eight: The cups are attached to the radiating flat splits at a distance to be determined by a tight succession of the sixteen cups around the circumference of the wheel (fig. 3). Each cup sits upon a radiating culm split. Attachment is achieved by drilling through the split into the walls of the cup securing with screws. Excess splits extending beyond the cups are now cut off at the cup. As a rough rule of thumb, the shorter height of the cups used will fit in a circle with a radius of slightly more than three times that height. For example, a 4" cup (short length and assuming a cup diameter of 3") would fit sixteen times in a circle with a radius of 14". With a central axle culm of also 3", the radiating culm splits would optimally be about 12". As the assembly progresses, care should be taken to align the cups such that they are in line, or perfectly concentric to the central axle.

Step Nine: To minimize frictional losses, the whole waterwheel should spin on the smallest diameter shaft that is sufficient to bear the weight of the wheel and water in operation. Also, the surface area of contact between the central shaft and the rotating wheel must be minimal. A simple system I have used is the

Figure 3

ADVANCED

following: using galvanized wire (18- to 20-gauge, wrap the wire according to the illustration (fig. 4) on either end of the culm central axle. A heavy gauge wire (coat hanger) is snaked through the bamboo axle and the wire bearing. Depending on the site of installation, the wire shaft can be supported by bamboo or other site specific support. The wire can be bent to prevent the wheel from sliding out of place (fig. 5).

Step Ten: This style wheel is an underflow type, the wheel rotates backwards. When finally situated in a stable and strategic position to catch the trough water, a final balancing must be performed. Unless you are extremely lucky, the center of gravity of the wheel will be eccentric to the central shaft. The wheel, when coming to a stop, will oscillate to one position- that which the center of gravity is directly below the shaft. To balance this, a weight of sufficient mass and distance from the axis of rotation must be installed to bring the center of gravity to the axis of rotation. When this is achieved, the wheel will spin and come to a stop randomly with no oscillation. The required counter weight mass is less if placed a greater distance from the axle. You may use a screw with a number of washers, finely tuning the weight to a balanced wheel.

Step Eleven: Finally, several measures can be taken to prevent or delay the degradation of the bamboo. Use an elastomeric roof coating to line the inside of the cups. It adheres to the bamboo, and makes a waterproof barrier capable of withstanding extreme dimensional variability (it stretches). To preserve the beauty of bamboo, coat all surfaces with a clear, water-based polyurethane. Allow the wheel to dry out, and operate the pump only when appreciated (parties, contemplation).

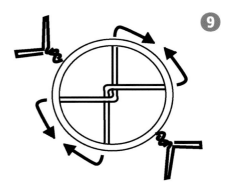

❧ *Figure 4* ❧

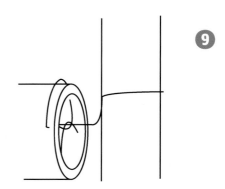

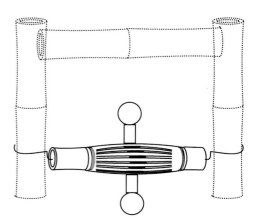

❧ *Figure 5* ❧

BURMESE BALL

"Chinlone" is a Burmese word referring to both the traditional ball and game that dates back to the seventh century A.D. Bamboo strips are woven wet; the weave tightens as it dries. Another game utilizing this design is a form of foot volleyball, popular throughout Southeast Asia. The design itself is a remarkable coming together of weave and form; six concentric rings weave perfectly to form a tightly fitted ball. It can be used for chandeliers and as a dome structure for climbing vines.

Materials

- *6 strips of 3'-long bamboo, flexible enough to loop into a circle*
- *Brass or galvanized wire or string*

Tools

- *Machete or Knife*
- *Vise Grips*
- *Wire Cutter*

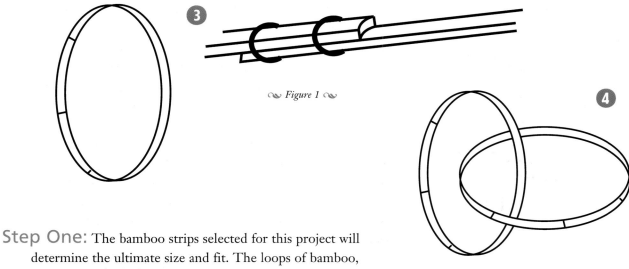

◈ Figure 1 ◈

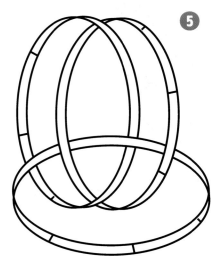

◈ Figure 2 ◈

Step One: The bamboo strips selected for this project will determine the ultimate size and fit. The loops of bamboo, woven correctly, are loose and difficult to manage until the fifth, and finally, the sixth loop is completed. It then suddenly tightens, and the form is a revelation. The illustrations accompanying the procedure are critical to getting the correct weave, so observe closely.

Step Two: Split all the 3' bamboo culms in half, then in $1/4$s, then $1/8$s.

Step Three: Make your first loop, and secure the overlap with wire or string. If the bamboo does not bend in a perfectly circular shape (likely), bend the bamboo in the areas where it is apparently stiffer, and shape the loop to your satisfaction (fig. 1).

Step Four: The second loop links with the first and is easily done (fig. 2).

Step Five: The third loop goes through both the previous loops and is also easily accomplished (fig. 3).

◈ Figure 3 ◈

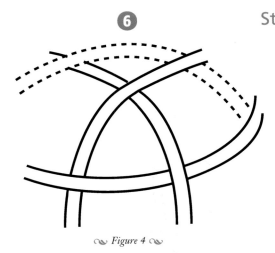

Figure 4

Step Six: The fourth loop requires a close inspection of the illustration (fig. 4). Arrange the three-loop ensemble such that they duplicate the orientation and weave arrangement shown. The weave of the bamboo ball is essentially a triangular weave; each loop forms one side of an equilateral triangle. The fourth loop essentially duplicates the loop on the opposite side of the crossed weave. The sequence of weaving into the three-loop arrangement is (over-under-under-under-over-over). The arrangement, at this point, is very loose, and must be held or secured to maintain the orientation and weave. If it does get away from you, it is possible to figure it out, and get it back to where it is supposed to be, but it involves some major mental calisthenics and is best to be avoided.

Step Seven: The fifth loop finally brings some structure and stability to the ball, relieving the anxiety and hassle of the previous step. Again, paying close attention to the illustration (fig. 5), weave the 5th loop accordingly. The sequence of weaving is (over-under-over-over-under-over-under-under).

Step Eight: The sixth loop is straightforward, the weave being the same ultimate weave of every other loop in the structure (over-under-over-under-over-under-over-under-over-under). And therein lies the beauty—six concentric loops, all woven over—under, to form a perfect ball—an awesome synergy of weave and geometry.

Figure 5

FENCE
ADVANCED

The best bamboo fence is live. It lasts forever, is self-replicating and is nice to look at. This is perhaps one of the most popular uses of bamboo for urban Americans. However, decorative bamboo fences have been raised to an art form in Japan, and there are many aesthetic arrangements of bamboo for garden accents. A basic, utilitarian fence is presented here. It is not completely bamboo, as the weavings and some horizontal members are integrated with a treated wood framework. Bamboo has a habit of deteriorating when exposed to the elements, but it can easily be replaced and the treated wood framework is relatively permanent.

Materials

- *Bamboo, larger diameter, for the vertical wood*
- *Bamboo for horizontal spans*
- *Treated 4 x 4 wood fence framework (The length of all your materials depends on your project needs.)*

Tools

- *Saw*
- *Machete/spoke shaver*
- *Drill and Drill Bit Set*
- *Hole Saw Kit*

Step One: Build the framework with treated wood as shown in figure 1. The height and number of weaves is subjective according to your design.

Step Two: The horizontal bamboo pieces are installed with complementary holes drilled into the vertical 4 by 4s. With one hole deeper than the other, the bamboo can be cut to a length allowing it to be inserted, spanning the span (fig. 2).

Step Three: Decide on the length of bamboo needed for the vertical weaves, and cut the bamboo culms making sure that neither end falls on a node.

Step Four: With a machete or spoke shaver, the bamboo culms are split only partially, leaving the last node and internode section intact. The culm is split in half, then in quarters. Turn the culm around, and do the same thing to the other end, only offset by 45 degrees, such that these splits are centered between the splits from the other end. The culm will stay together. Pick one split, and continue it through, such that the culm can be opened to a board. Knock off the remaining node septums. The bamboo board can now be woven into the fence.

84

Continue this process, tightly packing adjacent culm boards. Some pre-drilled screws may by needed to maintain the weavings in place. Especially if it is a loose weave. Alternatively, an additional weave can be added.

❶

∾ Figure 1 ∾

side view

❷

∾ Figure 2 ∾

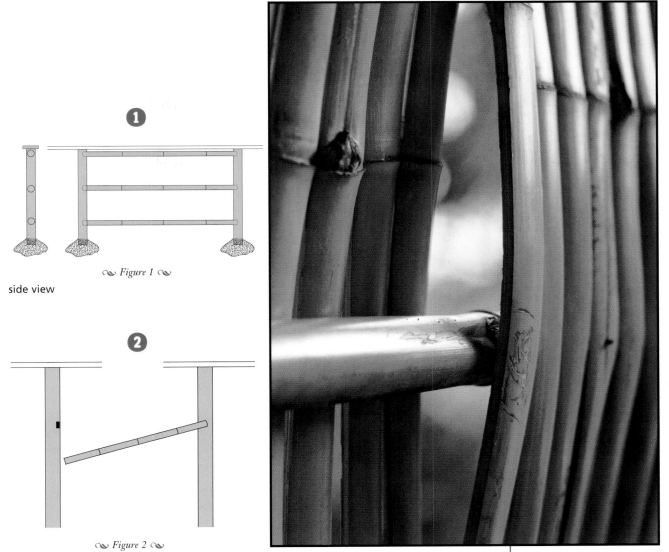

GARDEN GATE

*"Bamboo at my gate, to mend my mind," says an ancient Eastern proverb.
A strong and durable bamboo gate makes a fitting entrance. It sets the tone
for a relaxed, informal entry, and immediately encompasses the entrant
with the qualities of bamboo—light, smooth, round and friendly.*

Materials

- ❧ ½" I.D. galvanized pipe—48", 12", 24", and two 4" sections (standard lengths)
- ❧ 3 T-Joints
- ❧ 24' large-diameter bamboo
- ❧ 10' complementary smaller-diameter bamboo
- ❧ 16' of galvanized wire
- ❧ Nail or bamboo piece (to tighten wire)
- ❧ 36" by 2" by 6" bamboo

Tools

- ❧ Saw (mitre/pruning)
- ❧ Drill
- ❧ Hole saw kit
- ❧ Pipe Wrench/Vise Grips/Pliers
- ❧ Ground Digging Tools

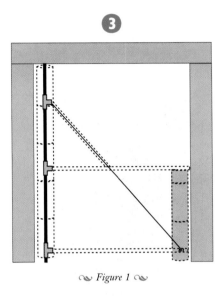

≪ Figure 1 ≫

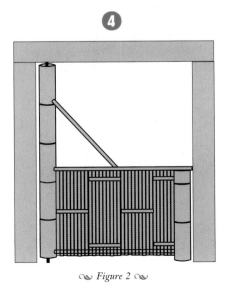

≪ Figure 2 ≫

Step One: Decide your gate's dimensions. The portal into which the bamboo gate is placed determines its size and dimensions. It would be advantageous to incorporate some shelter for the gate, as any protection from the elements would prolong the life of the bamboo gate. As a precaution, the horizontal surface of the gate may be made of wood and painted. This will serve as a cover for the woven bamboo and protect it as well.

Step Two: The bamboo is cut to fit around the pipe supports. Place ½" I.D. galvanized pipe, as the structural and pivotal internal components of the gate, hidden within the bamboo. The pipe will not deflect the beauty of the bamboo and will add considerable structural integrity.

Step Three: In addition, a double stranded wire diagonal support, also enclosed in bamboo, is used to prevent sagging. A diagram of the internal pipe and wire is shown in figure 1. The threaded pipe joints can be conveniently taken apart and reattached as needed.

Step Four: The largest-diameter bamboo is the vertical pivot and the outside vertical piece. Three complementary horizontal bamboos serve to allow the weaving of bamboo vertically. A sample gate is shown in figure 2. Note that the diagonal wire passes all the way through to the bottom horizontal to which it is attached. This ensures total gate support with the diagonal wire and puts all the bamboo in compression.

Step Five:
The diagonal bamboo, housing the double-stranded wire, is the key to fine tuning the gate, to make it square, strong and fit. The wire passes through the 2" by 6" and continues through the woven bamboo to attach to the outer end of the lower horizontal bamboo. A nail or another bamboo branch can be inserted between the two wires beneath the 2" by 6" piece of bamboo. There, it is equidistant between the two attachment points. The wire can be twisted and tightened until the desired configuration is attained (fig. 3).

Step Six:
Once all the components are pre-fitted, and a pre-assembly fit is satisfactory, the double wire is attached to the upper pipe, snaked through the diagonal bamboo support, through the 2" by 6", through to the lowest horizontal bamboo, inside the joint with the outside vertical bamboo as shown (fig. 4).

Step Seven:
The gate is installed by inserting the upper pipe into the upper guide and burying the lower pipe into the ground, positioned plumb. A final adjustment of the diagonal bamboo (turning it) will tighten any slack caused by weight and sag.

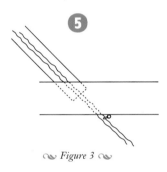

Figure 3

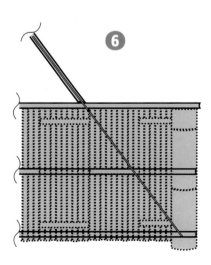

Figure 4

RESOURCES

AUSTRALIA

Bamboo Australia
 1171 Kenilworth Road, Belli Park via
 Eumundi, Queensland 4562, Australia.
 Phone and fax (61) 7 5447 0299
 bamboo@bamboo-oz.com.au
 www.bamboo-oz.com.au
 Bamboo poles—temperate and tropical varieties

EUROPE

Bambus-Centrum Deutschland
 D-75632 Baden-Baden, Germany
 Phone (49) 7221 5074 0
 Fax (49) 7221 5074 80
 info@bambus.de
 www.bambus.de
 Bamboo poles—imported, temperate varieties

La Bambouseraie de Prafrance
 Prafrance 30140 Anduze France
 Phone (33) 04 66 61 70 47
 Fax (33) 04 66 61 64 15
 bamboo@bambouseraie.fr
 www.bambou.bambouseraie.fr
 Bamboo poles—imported, various temperate
 varieties bamboo wares, crafts

JAPAN

Takehei Bamboo Wholesale
 403 Omiya-gojo, Shimogyo-ku, Kyoto
 600-8377 Japan
 Phone (81) 75 801-6453
 Fax (81) 75 802-6277
 takehei@mbox.kyoto-inet
 www.kyoyo-inet
 Bamboo poles—black, moso, other varieties

UNITED STATES

Bamboo & Rattan Works, Inc.
470 Oberlin Avenue S
Lakewood, NJ 08701
Phone (800) 422-6266
Fax (732) 905-8386
bam-booandrattan.com
Bamboo poles—tonkin, guadua, Taiwan reed fence

Bamboo Gardens of Washington
5016 192nd Place NE
Redmond, WA 98074
Phone (425) 868-5166
Fax (425) 868-5360
www.BambooGardenWA.com
Bamboo poles—tonkins, moso bamboo tools—
saw, knives, traditional twine

The Bamboo Man
7810 SW 118th Street
Miami, FL 33156
Phone (305) 378-9449
Fax (305) 378-2018
bamboo@bellsouth.net
Bamboo poles—tropical black, other clumping
varieties

Bamboo Supply Company
P.O. Box 5443
Lakeland, FL 33807
Phone (800) 568-9087
Fax (863) 646-8561
bamboosup@aol.com
www.bamboo-depot.com
Bamboo poles—tonkin, moso reed fencing

Big Bamboo Company
P. O. Box 1451
Dublin, GA 31040
Phone (912) 272-8544
A1bgbamboo@aol.com
www.bamboofencer.com
Bamboo poles—tonkin, yellow groove, black,
benon, giant timber, moso bamboo fencing,
fences

Frank's Cane & Rush Supply
7252 Heil Avenue
Huntington Beach, CA 92647
Phone (714) 847-0707
Fax (714) 843-5645
franks@franksupply.com
www.franksupply.com
Bamboo poles—black, various imported varieties

Ichiyo Art Center, Inc.

    432 East Paces Ferry Road

    Atlanta, GA 30305

    Phone (404) 233-1846

    Fax (404) 233-8012

    ichiyoart@aol.com

    www.ichiyoart.com

    Japanese shoji and decorative paper

The Japan Woodworker

    1731 Clement Avenue

    Alameda, CA 94501

    Phone (800) 537-7820

    support@japanwooderworker.com

    www.japanwoodworker.com

    Bamboo tools—Japanese saws, knives traditional twine

Smith & Fong Plyboo

    601 Grandview Dirve

    South San Francisco, CA 94080

    Phone (866) 835-9859

    Fax (650) 872-1185

    info@plyboo.com

    www.plyboo.com

    Bamboo flooring, plywood, paneling, veneers

Steve Ray's Bamboo Gardens

    250 Cedar Cliff Road

    Springville, AL 35146

    Phone (205) 594-3438

    www.thebamboogardens.com

    Bamboo poles—red margin, vivax, benon, Robert Young, other temperate varieties

Tradewinds Bamboo

    28446 Hunter Creek Loop

    Gold Beach, OR 97444

    Phone and fax (541) 247-0835

    gib@bamboodirect.com

    www.bamboodirect.com

    Bamboo poles—tonkin, moso, iron bamboo tools—Japanese saws, knives, splitters

Yucatan Bamboo

    5 Woods Edge Lane

    Houston, TX 77024

    Phone (713) 278-7344

    Fax (713) 278-7355

    Yucabambu@aol.com

    www.bamboofencer.com

    Bamboo poles—iron bamboo fences, gates, garden accessories furniture, interiors

INFORMATION

The American Bamboo Society
   750 Krumkill Road
   Albany, NY 12203
   Phone (518) 458-7625
   mab29@cornell.edu
   www.bamboo.org/abs
   Non-profit organization with global membership;
   provides source list of cultivated bamboo plants;
   publishes bi-monthly magazine and annual
   scientific journal

Temperate Bamboo Quarterly
   30 Myers Road
   Summertown, TN 38483
   Phone (931) 964-4151
   www.growit.com/bamboo
   Illustrated journals exploring the world of bamboo

FEATURED BAMBOO SITES

The Bamboo Farm and Costal Gardens
   2 Canebrake Road
   Savannah, GA 31419
   Phone (912) 921-5460
   Fax (912) 921-5890
   coastal@arches.uga.edu
   www.uga.edu/caes

Biltmore Estates
   One North Pack Square
   Asheville, NC 28801
   Phone (800) 543-2961
   www.biltmore.com

OUTDOOR APPLICATIONS

POLES

Bamboo Craftsman Co.
   2819 N Winchell
   Portland, OR 97217
   Phone (503) 285-5339
   www.bamboocraftsman.com
   Retail

Bamboo Gardner
   P.O. Box 17949
   Seattle, WA 98107
   Phone (206) 782-3490
   www.bamboogardner.com
   Retail

Bamboo Supply Company
   P.O. Box 5433
   Lakeland, FL 33807
   Phone (800) 568-9087
   Fax (863) 646-8561
   Retail

Livingreen
218 Helena Avenue
Santa Barbara, CA 93101
Phone (805) 966-1319
Fax (805) 966-1309
www.livingreen.com
Retail/wholesale

Steve Ray's Bamboo Gardens
250 Cedar Cliff Road
Springville, AL 35146
Phone (205) 594-3438
www.thebamboogardens.com
Retail

Takeroku America
Groton, MA
Phone (877) 692-3624
Fax (978) 433-4951
www.japanesebamboo.com
Retail

The Bamboo Man
7810 SW 118th Street
Miami, FL 33156
Phone (305) 378-9449
Fax (305) 378-2018
Retail

FENCES AND OUTDOOR STRUCTURES

Bamboo Fencer
179 Boylston Street
Jamaica Plain, MA 02130
Phone (800) 775-8641
Phone (617) 524-6137
Fax (617) 524-6100
www.bamboofencer.com
Fences, fencing; retail

Bamboo Giant Nursery
5601 Freedom Boulevard
Aptos, CA 95003
Phone (831) 687-0100
Fax (831) 687-0200
www.bamboogiant.com
Fencing; retail

Hammacher Schlemmer
Phone (800) 233-4800
www.hammacherschlemmer.com
Structures

Island Ambiance
459 Normal Avenue
Ashland, OR 97520
Phone (541) 482-6357
http://netdial.caribe.net/~bamboo
Treehouses

Safari Thatch & Bamboo, Inc.
  2036 C North Dixie Hwy
  Ft. Lauderdale, FL 33305
  Phone (954) 564-0059
  Fax (954) 564-7431
  www.safarithatch.com
  Structures and construction

TOOLS AND SUPPLIES

Bamboo Gardens of Washington
  5016 192nd Place
  Redmond, WA 98074
  Phone (425) 868-5166
  Fax (425) 868-5360
  www.BambooGardensWA.com
  Craft tools; retail/wholesale

Bamboo-Smiths
  P.O. Box 1801
  Nevada City, CA 95959
  Phone (530) 292-9449
  Fax (530) 292-9460
  tbs@sierratimberframers.com
  Saws, splitting and carving knives, miscellaneous
  supplies; retails

Hida Tools
  1333 San Pablo Avenue
  Berkeley, CA 94702
  Phone (800) 443-5512
  Phone (510) 524-3700
  Fax (510) 524-3423
  www.hidatool.com
  Saws, knives, splitters, bits; retail

Hirade America
  51 Shattuck Avenue
  Pepperell, MA 01463
  Phone (877) 692-3624
  Fax (978) 433-4951
  www.japanesetools.com
  Bamboo woodworking tools from Japan—saws,
  chisels, splitters, thread makers

Misugi Designs
  Phone (707) 422-0734
  Fax (707) 425-2465
  www.misugidesigns.com
  Japanese woodworking tools; retail

The Japan Woodworker
  1731 Clement Avenue
  Alameda, CA 94501
  Phone (800) 537-7820
  www.japanwoodworker.com
  Japanese saws, knives, splitters. rope; retail